An Introduction to Reptiles and Amphibians For All Ages

James Vanas

Copyright © 2023 by – James Vanas – All Rights Reserved.

For my grandson Gus, who loves the natural world and helped to author this book

INTRODUCTION

The huge and powerful jaws of a crocodile, the long sharp fangs of a venomous snake, and the highly poisonous skin of a tiny dart frog may make you feel uncomfortable about the world of reptiles and amphibians. But the truth is that most of these animals are harmless critters that help keep our world in balance in a meaningful way. Using common sense around places where dangerous reptiles and amphibians exist will keep you safe.

Their world is fascinating and diverse and sometimes hard to understand. How can a frog in the north freeze and thaw as spring arrives? How does an alligator provide water during a drought that saves the lives of countless birds, frogs, fish, snakes, and mammals? Knowing about these animals will give you a greater appreciation of the world around you.

This book is designed to give you a general look at the six scientific orders of reptiles and amphibians. One thing is certain: within their natural histories, there are more exceptions to what we think should be a standard rule or trait.

My grandson, Gus, became interested in this project and contributed by writing a page for each chapter and adding an original illustration. It has been a great experience for both of us. The last page of each chapter contains some informative yet whimsical poems I have written.

Rhinoceros Viper *Bitis nasicornis*
Photo: Seth Cohen

I have been involved with the study of reptiles and amphibians for over 50 years as an amateur scientist. You do not need to be a trained professional to enjoy the world around you. I have tagged sea turtles in Florida, moved nuisance alligators, worked with giant reptiles at a private zoo, and most recently been involved in the Mojave Desert tracking Gila monsters and tortoises.

So, let's get started. I hope it is fun and exciting. Above all, I hope it helps you better understand the crazy, extraordinary world of these iconic creatures.

TABLE OF CONTENTS

Understanding Scientific Names
Chapter 1 Salamanders and Caecilians — 01
Chapter 2 Frogs and Toads — 09
Chapter 3 Snakes — 17
Chapter 4 Crocodiles and Alligators — 25
Chapter 5 Lizards — 33
Chapter 6 Turtles — 41

Bonus Pages
Sexual Dimorphism Explained — 49
Florida's Exotic Nightmare — 50
Reptiles Maximum Sizes and Ages — 52
Reptiles and Amphibians that We Eat — 53
Reptiles and Amphibians As Pets — 54
Reptiles and Amphibians By The Numbers — 55
New World and Old World Explained — 56
Field Guides — 57
Venom Vs. Poison — 58

Conservation efforts — 59
Acknowedgements — 60
About the Author — 62

SCIENTIFIC AND COMMON NAMES

Throughout the book, you will see both common and scientific names used. This page will hopefully give you an understanding of the concept of scientific names and why they are important.

TAXONOMY is the science of naming, defining, and classifying groups of living organisms. To keep things simple in this book, the **Genus and Species** names will be used. In some cases, the **Order** and **Family** of animals are mentioned.

The idea of scientific names can appear overwhelming. It is important in zoology to have a name for an animal that scientists all over the world can recognize. Common names are just too confusing. Like written music, scientific names are a universal language.
Genus - is the first of the two names used to describe a similar group of animals.
In print, this name is always Capitalized and italicized.
Species - is the second name and is used to describe a specific animal within the genus. In print, the species is written in lowercase letters and is italicized.

For example, many kinds of rattlesnakes are found in the Genus Crotalus.
Crotalus atrox - The Western Diamondback Rattlesnake.
Crotalus adamanteus - The Eastern Diamondback Rattlesnake.

Subspecies - an animal that is the **same species** but may look a little different or be found in a different location.
For example, Gila monsters are known to science as *Heloderma suspectum*. In its northern range, it is named *Heloderma suspectum cinctum*, commonly called a banded Gila. In its southern range, it is named *Heloderma suspectum suspectum*, and commonly called the reticulated Gila.

Banded Gila monster

Reticulated Gila monster

A species name can be very descriptive. It can describe **a place** where the animal is found or **a particular trait** of an animal. It may honor the **name of the person** who discovered the animal or the **name of a person that scientists wish to honor**.
Crotalus catalinensis - A small rattlesnake found on Isla Santa Catalina: a small island off the coast of Baja, Mexico.
Crotalus mitchelli - The Speckled Rattlesnake was named for a doctor, S.W. Mitchell, who studied rattlesnake venom.
Crotalus ruber - The Red Diamond Rattlesnake. Ruber is a Latin word meaning "red".
Some scientific names can be very simple to read and pronounce. For example, the boa constrictor is known as the **Boa constrictor**. The Green iguana is *Iguana iguana*.

SALAMANDERS

There are 700 plus species of salamanders
Order: Caudata

and CAECILIANS

There are 200 plus species of caecilians
Order: Gymnophiona

Western Tiger Salamander *Ambystoma mavortium diaboli*

Axolotl *Ambystoma mexicanus*

Crocodile salamander *Tylotoriton shanjing*

Lesser Siren *Siren intermedia*

a Caecilian species

Salamanders make up only 9% of all amphibians, while Caecilians only 3%. All the rest are frogs and toads. More species of both orders are still being discovered.

Salamaders have the ability to regenerate limbs, tails, internal organs and even brain tissue. The majority of salamanders are lungless and breath through their skin.

SALAMANDERS

Salamanders look a lot like lizards, with four legs, a long tail, and a slender body. However, within this small order of amphibians, there are many exceptions in their appearance, lifestyle, reproductive and respiratory strategies. Most salamanders are small, averaging 2 to 6 inches. The largest is the Chinese Giant Salamander, Andreas davidianus, which can achieve a total length of 5 feet.

Although salamanders are found in Central and South America, Europe, and Asia, **they reach their greatest diversity in the Appalachian Mountains** of the eastern United States, where 77 species are recorded. In comparison, only 30 species have been named in all of South America.

Some salamanders live on dry land (terrestrial), some underground (fossorial), a few live in trees (arboreal), and others never leave the water (aquatic), but most are considered semi-aquatic with a life cycle similar to that of frogs. Eggs are laid in water and hatch into larvae with gills, followed by a juvenile stage and adulthood.

Eastern Tiger Salamander
Ambystoma tigrinum

Regardless of where they live, salamanders have moist, usually smooth skin that requires them to avoid temperature extremes and stay near water or stay hidden under logs and debris in humid woodlands. Most salamanders have poison glands on their bodies and tails. These toxins discourage predators from eating them. Although the poison must be ingested to be harmful, handling salamanders should be done with protective gloves. It is better not to handle them at all.

Salamanders can regenerate lost tails, much like lizards. They can also grow new legs, and some can regenerate internal body organs and parts of their spinal cord. There seems to be nothing typical about salamanders, which makes them all the more fascinating.

CAECILIANS

Striped Caecilian
Ichthyophis supachaii

Caecilians are limbless, worm-shaped amphibians that live in Central and South America, southern Asia, and Africa. They spend most of their lives underground in tunnels and burrow in soft soil.

A few are totally aquatic. With this cryptic lifestyle, they are rarely seen, and little is known about them. Some may grow to 5 feet long, but most are much smaller. They have rings that encircle their body and sensitive tentacles between their nostrils that help them find food. Insects, worms, mollusks, snakes, frogs, and lizards are eaten. With strong jaws and very sharp teeth, some large caecilians may even eat small mammals. The smooth skin of caecilians is toxic. Some species bear live young, and others lay eggs and develop much in the same way as salamanders.

SALAMANDERS
Some examples

THE EASTERN TIGER SALAMANDER: Ambystoma tigrinum is the largest terrestrial salamander in North America. They average 6 to 9 inches in length but have been recorded reaching 13 inches. Their geographic range extends from Alaska and Canada south through much of central and eastern United States and into Mexico.

They are part of a family known as "mole salamanders" and **spend most of their time buried underground.** In late winter and early spring they will migrate, sometimes in large numbers and long distances, to vernal ponds to breed. A vernal pond, or wetland, is void of predatory fish and goes dry each summer.

THE FIRE SALAMANDER: *Salamandra salamandra* is a European species with vibrant orange or yellow markings on a black background. **They have poison glands behind their eyes and can squirt a stream into the face of a would-be predator from a foot away.** These 6-12 inch long beautiful salamanders are lungless and breathe through their skin. Unlike most salamanders, the female will retain eggs inside her body, and then place formed larvae in the water to further develop. Commonly kept as pets, the fire salamander has been known to live for 30 years.

Fire Salamander
Salamandra salamandra

SIRENS: These very strange salamanders are aquatic and paedomorphic, (retaining larval features as an adult). Sirens look like eels but have external gills and two small, useless front legs. The greater siren, *Siren lacertina*, can grow close to 3 feet long and is found in swamps, ponds, lakes, and drainage ditches in the southeastern United States. They are nocturnal and feed on insects, crayfish, snails, and fish. They can survive a drought for months or years by burrowing in the mud and forming a mucus-filled cocoon around their body.

Red-spotted Newt
Notophthalmus viridescens

EASTERN RED SPOTTED NEWT: All newts are salamanders. Newts just have a little different life path. After hatching from eggs, the larva develops and crawls out of the water, looking like other salamanders but with rough, warty skin. **In this juvenile stage, they are called an eft.** For several years, they stay in this condition before moving back into the water, where they lose their color and live out their adult life. The Red-spotted newt is very common and found throughout the eastern half of the United States.

featured creature
THE HELLBENDER
Cryptobranchus alleganiensis

DESCRIPTION: This fully aquatic beast is the largest salamander in the U.S. They have a flat body and round head, slimy skin, short legs, and small beady eyes. The hellbender is reddish brown in color with dark blotches on top. They have a paddle-shaped tail. Adults are usually around 20 inches long and can weigh 4-5 pounds. The flattened, thick folds of skin on their sides look a bit like lasagna. Oxygen is absorbed through these skin folds. Adult hellbenders have no external gills and no eyelids.

RANGE: Hellbenders are found from southern New York to Georgia and Alabama along the Appalachian Mountain chain. They are also found in Ohio, Indiana, Illinois, and Kentucky. An endangered subspecies, the Ozark hellbender, is found in Missouri and Arkansas.

DIET: **Hellbenders feed mainly on crayfish** but will eat snails, worms, insects, fish, and other hellbenders. Their eyes are weak, but their sense of smell is keen. With their large gaping mouth, they can swallow an animal close to their own size.

HABITAT: Hellbenders are habitat specialists. They require clear, fast-moving, highly oxygenated, pollution-free streams with large flat rocks to hide under. If the water becomes silted with sand and mud, they die. These high-quality streams are becoming rare.

REPRODUCTION: Male hellbenders fight viscously in late summer as the mating season heats up. The winning male excavates a small saucer-like depression under a flat rock and waits for females to come and lay between 200-400 eggs. He fertilizes them from above her and then chases her away. **The male protects the egg mass** from predators for 50 days or more until they hatch. The young salamanders are on their own after hatching and will become mature hellbenders in 5-6 years. Adults may live for 15 years or more.

featured creature
THE HELLBENDER

Finding hellbenders can be quite difficult. When wading in streams, some people turn over large flat rocks while looking for them. A dropped rock can crush the salamander. Walking in a stream can disturb the egg mass a male is guarding. However, the major threat to their survival is stream pollution from agricultural chemicals, construction projects, and the damming of rivers that silt and slow the necessary flow and depth of the water. It is said that **the hellbender population has decreased by 80%** over their range since the 1970s.

Conservationists are trying to increase the number of hellbenders. **Headstart Programs** conducted by government agencies with the cooperation of many zoological parks, remove a few egg masses and raise the salamanders for a number of years before releasing them back into the wild.

Photo: Kevin Stohlgren

Herpetologists trying to find the secretive hellbender in streams take water samples and test for DNA, which can indicate whether or not they are present. This **eDNA** (environmental DNA) trick can save countless hours of searching and help to identify streams that need protection.

Hellbenders have lots of nicknames. The **snot-otter, devil dog,** and **lasagna lizard** are a few, but no one seems sure how the name hellbender came into use. Some claim that early settlers, upon first seeing them, said, "They look like they escaped from hell and are **bent** on getting away."

These ancient, nocturnal salamanders are the largest amphibians in the United States. They look very similar to the huge Japanese and Chinese Giant Salamanders that grow to extraordinary lengths of 5 feet or more.

GUS' PAGE

THE CHINESE GIANT SALAMANDER

Andrias davidianus

DESCRIPTION: The giant salamander has smooth skin and lots of folds for getting the oxygen they need. Their one lung is used to control buoyancy. They have 4 stubby legs and tiny eyes, and the tail makes up half of their body length. They may grow close to 6 feet long and weigh 55 pounds. They are the largest salamander in the world.

RANGE: They are found in central and southern China, but their range has been limited due to the destruction of their habitat and collection of food and medicine. There are captive breeding programs in place.

DIET: They eat crayfish, fish, insects, worms, frogs and snails. They may also eat other smaller salamanders of their own kind.

HABITAT: Fast-flowing and cold streams are where these salamanders live. They like to hide under large rocks and are hard to find.

REPRODUCTION: Females will lay 400-500 eggs in a cavity where the male will fertilize them and protect them until they hatch in 30-60 days. The young will be on their own when they hatch and will be adults in 5-6 years.

FUN FACTS and EXCEPTIONS SALAMANDER

Mudpuppy, or waterdogs, are aquatic salamanders that never leave the water and remain in their larval state. This condition is referred to as **neoteny**. The common mudpuppy of the central United States, Necturus maculosus, grows to be nearly a foot long and lives in rivers, streams, and lakes. Fishermen consider them the best bait for bass.

Amphiumas are elongated aquatic salamanders that look like sirens. They can grow to be over 3 feet long. Their four legs are tiny and useless. They have no eyelids or tongues. They do not have external gills. They live in the southeastern United States, and like the siren, they are hard to find.

A rare Costa Rican salamander
Oedipina carablanca

Most salamanders reproduce internally. The male deposits a plug of sperm, called a **spermatophore,** in the water. The female picks up the plug with her **cloaca** (the vent that opens to the intestinal, urinary, and genital tract). After laying her eggs, she may then guard the brood until they hatch. A notable exception is the external fertilization that occurs with hellbenders.

A few salamanders have tongues 10 times longer than their body. When salamander eggs hatch, they are called **larvae**, not tadpoles.

Ambystoma mexicanum

The rare, critically endangered **Mexican axolotl** is a salamander that stays in its larval form throughout life. It has feathery gills, four legs, and a paddle-shaped tail.

It is found in only a few lakes near Mexico City. Females may lay upwards of 1000 eggs, which are left on their own to hatch and mature. The average lifespan of a wild axolotl is about 5 years.

Like some other salamanders, the axolotl can regenerate limbs. However, it can also **regenerate parts of its head and spinal cord, its heart, and even its eye**. All of this is done without any scars, and once healed, it can regenerate the same organ or limb many times over.

Axolotls breed well in captivity, and many are kept as pets around the world. They are usually brown in color, but albino forms are popular in the pet trade. Non-native fish in Lake Xochimilco and pollution are driving them close to extinction in the wild.

Salamander Poems

Regeneration

So you bit off my tail but that's okay
I'll grow a new one for another day.

Then you bit off my leg and that's not cool
I'll grow that back-that's how I rule

And even if you bite through my skin
and damage the organs I hold within

I'll figure a way to refit myself
and return to a life of everyday health.

Regeneration is what I can do
we salamanders have a lot to teach you

Alpine Newt *Mesotriton alpestris*

Three-lined Salamander
Eurycea guttolineata
Photo: Cameron Rognan

Searching for Salamanders

The salamanders soft moist skin
makes him seek protection
under logs and under rocks
he avoids detection

Mistaken for a lizard
which actually he's not
amphibian is the name
given to his lot

Related more to slimy frogs
and to the bumpy toad
if no one ever told you
then how would you have know'd

In searching for this little beast
plan to go meandering
in the woods and near the streams
when you go "salamandering"

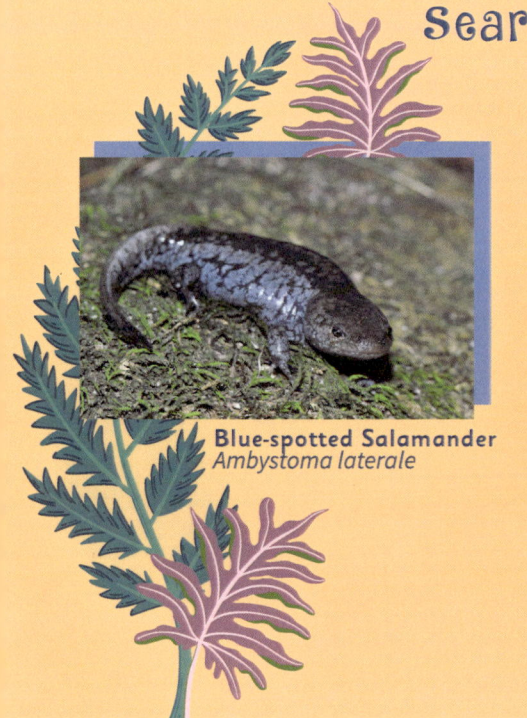

Blue-spotted Salamander
Ambystoma laterale

FROGS AND TOADS

There are around **7,000** species of frogs in the world
Order: Anura

Lemur Leaf Frog *Agalychnis lemur*

Coronated Tree Frog *Triprion spinosa*

Glass Frog *Centrolenella ilex*

Strawberry Dart Frog *Oophaga pumillo*

FROGS and TOADS

Frogs and toads are found on every continent except Antarctica. They can live in almost any habitat, from low, hot deserts to high, cold mountains and humid tropical rainforests.

This extremely diverse order includes frogs that are active during the day (diurnal), at night (nocturnal), in trees (arboreal), on the ground (terrestrial), and in the water (aquatic). They may be brilliantly colored or rather drab. Some frogs are very vocal. Male frogs croak considerably more than females.

Frogs are **carnivores** that provide a great service by helping to control insect populations. With sticky tongues, they catch their prey, and if it fits in their mouth, they eat it.

Herpetologists group frogs and toads together in the order Anura as they are really the same, with just a few minor differences:

Spadefoot Toad

Red-eyed Tree Frog

• Frogs have **thin, smooth**, and somewhat **slimy skin**.

• Frog **legs are long** and made for **jumping**, some for very long distances.

• Many frogs have **tiny teeth**-like structures on their upper jaw for holding prey.

• Most poison dart frogs are **not very dangerous**, but one species is deadly if handled.

• Toads have **thick, bumpy skin** and can travel farther from water.

• Toads have **short legs** and **hop** rather than jump.

• Toads have a **parotid gland** behind the eye that can exude milky, sticky poisons.

Both frogs and toads need to be **near water** to complete their interesting **life cycle**.

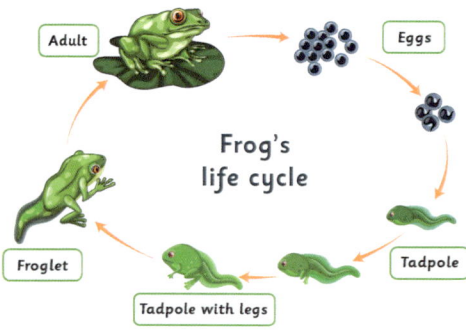

Frog's life cycle

Frogs lay jelly-like eggs, usually in water after mating. Eggs hatch into tadpoles, which eventually transform (metamorphose) by losing their tails and gills, developing lungs, and growing legs. There are some exceptions, but this is the general process.

Frogs suffer from humans disturbing and destroying valuable wetland habitats. Around the world, a devastating infectious disease called **Chytrid fungus**, or **Bd**, has wiped out populations and caused some extinctions. No cure has yet been found for Chytrid.

FROGS and TOADS
Some examples

TREE FROGS: About 800 species of tree frogs (family *Hylidae*) are found living around the world. Most, around 600 species, are found in the western hemisphere and largely in the tropics. Not all tree frogs are arboreal, but they all have a claw-shaped bone on their last toe, which helps them cling on to branches wherever they are. Most tree frogs are small and green and blend in well with the environment. Male tree frogs make "advertisement calls" during breeding season to attract mates.

GLASS FROGS: These small green frogs live in the Central and South American lowland rainforests high in the treetops. **Several species of glass frogs have translucent skin on their bellies, making their internal organs, such as the heart, liver, and intestines, visible.**

During breeding season, a female deposits eggs on top of a leaf above a stream or pond. The male then fertilizes the eggs and stays with them day after day until they hatch and fall into the water as tadpoles.

THE CANE TOAD: *Rhinella marina* This large (avg. 4-6 inches) toad is native to the Americas from southern Texas to the Amazon. From there it has been introduced around the world and is an INVASIVE NIGHTMARE.

In the mid-1930s, around 100 of these **highly poisonous toads** were shipped to Australia to eat cane beetles and other pests in the sugar fields. After the fields were clean, the toad started eating everything, including insects, snakes, other frogs, lizards, and even small mammals. Wild and domestic animals that eat them get sick and may die.

They have spread across the country and their numbers have exploded to an estimated population of 250 million plus. With no effective control, they are devastating native wildlife species.

In Florida, they are referred to as "Bufo or Marine" toads. They are also found in Puerto Rico and are an invasive pest.

featured creatures
POISON DART FROGS

DESCRIPTION: Poison dart frogs are small, brightly colored amphibians. The average size is between 1-2 inches. The **golden poison dart frog**, found only in Colombia, is one of the most toxic animals on earth, with enough poison on its skin to kill 6 humans. The majority of dart frogs are far less toxic.

Golden Poison Dart Frog
Phyllobates terribilis

RANGE: Poison Dart Frogs are found in humid tropical areas of Central and South America from Nicaragua to Brazil.

HABITAT: These frogs are mainly terrestrial and semi-arboreal and may be found in areas of degraded pastures, grasslands, rural gardens, and pristine lowland tropical rainforests. They are diurnal hunters and may be very vocal.

Dart Frog with tadpoles
Hyloxalus azureiventris

DIET: Dart frogs are insectivores that eat ants, beetles, termites, and other small insects. Scientists think that the poisons they secrete are a result of the food they consume. In captivity, frogs fed crickets, worms, and fruit flies lose their poison potential and are harmless when handled.

REPRODUCTION: Male dart frogs lead a female to a moist place on the ground and externally fertilize the eggs she lays. After the eggs hatch, the female, and sometimes the male, gather the tadpoles on their backs and carry them to a permanent water source where they develop into adults. They are good parents and very protective of their tadpoles.

Harlequin Dart Frog
Oophaga histrionica

featured creature
STRAWBERRY DART FROG

Oophaga pumilio

DESCRIPTION: The strawberry poison dart frog is a small, slender frog less than one inch long. Their bright red dorsal color is common, but colors can vary widely throughout their range in Nicaragua, Costa Rica, and Panama. As many as 20 different color types or "morphs" are known. Some may be black, yellow, orange, or blue, with many dark spots, blotches, and crossbars. One strikingly beautiful color morph is called the Blue Jeans Frog. They have a bright red body and dark blue legs. Bright colors warn potential predators of their poisonous nature.

A FASCINATING AND UNIQUE LIFE CYCLE:

• The male frog will lead the female to a wet leaf or depression on the forest floor.

• There, she will lay 5 or more eggs, and he will externally fertilize them.

• He will protect them, clean off any fungus, rotate them, and keep them moist.

• When they hatch, the female will return and load the tadpoles onto her back.

• She will climb high into the canopy, perhaps 50 feet up, and deposit the eggs into a water-filled tank bromeliad. To avoid cannibalism, she places only one egg per plant.

• Then, amazingly, she will return to each tadpole every few days for 6-8 weeks and lay an unfertilized egg for them to eat. Without this food, they could starve.

• As the tadpoles sprout legs and mature, they climb back down to the forest floor.

CAN YOU IMAGINE A ONE-INCH FROG DOING ALL THIS?
The Genus name, Oophaga, means egg-eater, which is what the tadpoles do.
The species name, pumilio, means dwarf, which is what these frogs are.

Strawberry Dart Frog "Blue Jeans"

Strawberry Dart Frog with tadpole on back

Strawberry Dart Frog color morph

GUS' PAGE

THE AMERICAN BULLFROG
Lithobates catesbeianus

No bullfrogs where I live so I am holding a toad

DESCRIPTION: Their color can be brownish to different shades of green. They can have dark spots along the back. Bullfrogs can get to about 6 inches and can weigh up to 1 pound or more. They are the largest true frogs in North America.

RANGE: Originally, they were found in the eastern United States and SE Canada. Bullfrogs were taken to places like California and Colorado. They were also exported to southern Europe, South America, and Asia to frog farms to raise them for food. Escaped bullfrogs now compete with native species.

DIET: Bullfrogs can eat anything they can fit in their mouths, including snakes, worms, insects, crayfish, salamanders, and other frogs. They can be cannibalistic, meaning they will eat smaller bullfrogs.

HABITAT: Bullfrogs must live near water. They are usually seen near lakes, ponds, swamps, and bogs. They prefer warm, shallow water and hibernate when the weather gets cold. Bullfrogs can live to 9 years in the wild. People hunt bullfrogs for their legs.

REPRODUCTION: Fertilization is external. Females can lay 20,000 eggs. After 4 or 5 days, the tadpoles hatch and are on their own. Most tadpoles will be eaten by a variety of predators.

FUN FACTS and EXCEPTIONS FROGS

Almost all frogs reproduce in the same way. Males fertilize the female externally as she lays her eggs. Eggs in the water hatch into tadpoles, which grow and eventually metamorphose, in an amazing way, into little frogs. Almost every organ in the tadpole has to change in order to become a froglet.

Surinam Toad *Pipa pipa*

There are exceptions to this process. The tailed frog, Ascaphus truei of the Pacific northwest United States, is a voiceless two-inch amphibian that lives in cold, fast and clear streams. The tail of the male is actually an extension of the cloaca (the vent opening for digestive, reproductive, and urinary functions). The frog uses this "tail" to internally fertilize the female in order to keep the sperm from floating away. The female must attach her eggs to rocks or sticks underwater so they are not swept away.

The Surinam Toad is an aquatic frog from South America that is popular in the pet trade. It is commonly called by its scientific name, **Pipa pipa**. The female lays eggs outside the body, and the male fertilizes them as usual. He then proceeds to push the eggs onto the female's back, where small pouches form around each egg. The eggs develop into froglets that burst out of her back and go on their way.

Frogs live in so many diverse habitats that it is important for them to be able to adapt and survive. In northern climatic zones, some terrestrial wood frogs actually freeze solid and thaw out as the weather warms. Aquatic frogs may spend the winter under frozen ponds just sitting on top of the mud.

The smallest vertebrate in the world is a frog. A vertebrate is an animal with a backbone. From Papua New Guinea, the tiny frog, **Paedophryne amanuensis**, is only one-quarter inch long. The Goliath frog, **Conraua goliath**, from Western Africa, is a monster. They can grow to be a foot long and weigh seven pounds. Goliath frogs can jump almost 10 feet.

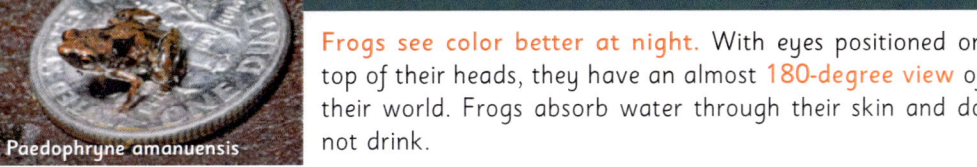
Paedophryne amanuensis

Frogs see color better at night. With eyes positioned on top of their heads, they have an almost 180-degree view of their world. Frogs absorb water through their skin and do not drink.

Some desert frogs in arid areas of Australia dig deep burrows when a drought comes and encase themselves in a mucus-filled cacoon, where they can survive for years before it rains again.

In the United States, there are 100 species of frogs. In Brazil, there are 700 species. Frogs are more diverse in the warm tropics.

FROG POEMS

Frogs Live Everywhere

" Frogs a jumpin' and flying frogs
Frogs in the desert and frogs in the bogs.

Frogs in the classroom on dissecting tables
Frogs in jars with informative labels.

Frogs in the tropics that live in the trees
Frogs in Alaska surviving a freeze.

Frogs with bright colors and poisonous skin
Frogs that are huge or as small as a pin.

Frogs are our friends coz they eat lots of bugs
Frogs can be cute but frogs don't give hugs.

Frogs are a food source you can catch with a permit
Frog legs are scrumptious but please don't tell Kermit. "

Poison Dart Frogs

" If you think you want to pick me up
You'd better think again.
I may be small and pretty BUT
There's poison on my skin

Bright colors warn my enemies
I'm not what they desire
And if you choose to catch me
I may set your soul on fire.

I'm the jewel of the tropics
A rare and awesome frog
So look at me but let me be
Go home and pet your dog "

Yellow Banded Dart Frog
Dendrobates leucomelas

SNAKES

There are over 3,800 species of snakes worldwide
Order: Squamata

Black-tailed Rattlesnake *Crotalus molossus*
Photo: Cameron Rognan

Rosy Boa *Lichanura trivgata*

Eyelash Viper *Bothriechis schlegelii*

Mexico has the most species of snakes in the world with 428. Brazil is second on the list, followed by Indonesia and India.

Mexico also has more venomous snakes than any other country.

The United States is home to about 170 snake species, with only 30 being venomous.

Ringneck Snake *Diadophis punctatus regalis*
Photo: Seth Cohen

Coral Snake *Micrurus alleni*

SNAKES

In nature, snakes are easy to recognize. With legless, elongated bodies covered with overlapping scales, these carnivorous (meat-eating) reptiles are known to all of us.

Snakes lack moveable eyelids and external ear openings. With forked tongues, they rely on "smelling and tasting" the environment, using the sensory Jacobson's Organ on the roof of their mouth. The snake flicks its tongue in the air picking up airborne particles, and then retracts the tongue inside the mouth and touches it to the organ to identify chemical clues as to where a prey animal may be.

Panamint Rattlesnake
Crotalus stephensi

Some snakes are spaghetti-thin, look like worms, and are less than 4 inches long. Pythons and anacondas are huge. A female anaconda from South America may exceed 20 feet in length and weigh 350 pounds.

A snake's diet depends on their size. Many smaller species eat insects, small lizards, and little fish. There are snakes that eat snails and snakes that eat the eggs of tree frogs.

Some snakes specialize in eating scorpions. Small rodents, birds and their eggs, and frogs make up the diet of many snakes. Pythons, boa constrictors, and anacondas can consume large animals like deer, capybaras, alligators, and cayman.

There is no doubt that snakes are extremely important to us as they help to keep populations of pests, like disease-spreading rats and mice, in check. Therefore, killing a snake because you assume it is venomous is not a logical act. Learning to identify snakes where you live and protecting them from harm makes more sense.

Snakes are solitary animals, but in colder climates, they may often den together in large numbers in what is called a hibernacula. When spring arrives, snakes may mate before they spread out into the environment. This is a big advantage and saves the snakes a lot of energy and time. Some rattlesnakes and garter snakes are known to den in large numbers.

Not all snakes lay eggs. Some bear live young, like water snakes and rattlesnakes. A hatchling snake from an egg looks like a miniature copy of the adult in most cases.

SNAKES
Some examples

PYTHONS AND BOA CONSTRICTORS: Pythons are "Old World" snakes from Africa, Asia and Australia. Boas are primarily of the "New World," North, Central, and South America. The family of boas includes anacondas.

Pythons and boas are constrictors and ambush predators. Once a prey animal is captured in its jaws, these snakes wrap coils around it and keep squeezing. The animal may suffocate, or some scientists think the circulatory system fails and the animal dies. Either way, prey is then swallowed head first and whole. They are not bone crushers, as some people may think.

Some boas and pythons have heat-sensing pit organs on their jaws that help them to accurately detect warm-blooded prey even on the darkest night.

Boa Constrictor

We think of pythons being large snakes, and most of them are. However, among the 33 or so species in this family, the pygmy python from western Australia achieves a length of only 23 inches. A big reticulated python of Southeast Asia may grow to be over 20 feet long.

Pythons lay eggs, and female pythons may wrap around the egg mass in order to keep them warm as they incubate. When hatched, the young are on their own. **Boa constrictors and the anaconda bear live young.**

Pythons are **not good pets**. A small python in captivity can grow huge in a matter of a few short years. Unfortunately, in the Florida Everglades, a breeding population of Burmese pythons now exists, as pet owners released their snakes over the years when they became too much to handle. Without natural enemies these snakes are eating everything and are damaging the environment in significant ways. Python hunts encourage citizens to catch and destroy these snakes in South Florida.

THE INDIGO SNAKE *Drymarchon couperi*: The beautiful, glossy, and iridescent indigo snake of the Southeast United States is the longest native snake in North America. They may average 5-6.5 feet but have been recorded reaching over 8 feet in length. These gentle giants rarely bite when handled, but they are tough and aggressive predators that are **known to eat rattlesnakes.**

Indigos are found in Florida and Georgia and are protected by law. They are no longer found in Alabama or Mississippi. Habitat destruction of pine flat woods and over-collection for the pet trade have made this amazing snake a rare reptile.

featured creatures
RATTLESNAKES

There are 36 species of rattlesnakes found in North, Central, and South America. All are venomous. Most rattlesnakes are, on average, 3 feet long or smaller. A few big ones, like the Eastern and Western Diamondbacks, may exceed 5 feet in length and are dangerous reptiles. **KEEP YOUR DISTANCE AND YOU WILL BE SAFE**.

Eastern Diamondback Rattlesnake
Crotalus adamanteus

Rattlesnakes are pit vipers. On their upper jaw, between the nostrils and the eyes, are two pit organs that are infrared sensitive and help rattlesnakes detect minor temperature differences of nearby prey animals. Even in total darkness, a rattler can "see" a warm-blooded mammal that comes within striking distance.

Known as an **ambush predator,** a rattlesnake will wait near a mammal trail and strike an unsuspecting mouse, rat, or rabbit as it passes. The prey is instantly released, which keeps the snake from being injured by a mammal's sharp teeth.

The venom works quickly, and the snake then uses its forked tongue and amazing sensory organs inside the mouth to track its path and swallow the dead prey whole.

Though rattlesnake venom can kill, it can also heal. **Today, many medications are made from their venom.** Pain medicine, blood pressure, and heart attack medicines use snake venom as part of their cocktail mix. Cancer medicines are being tested.

For the snake, venom is a valuable resource not to be wasted. Rattlers can actually control the amount of venom they deliver when striking. **It is reported that in 25% of bites to humans, no venom is delivered.** These are known as "dry bites."

As snakes grow, they shed their skin in one long piece, starting at the head. Rattlesnakes retain part of the shed, which adds to their tail rattle. These rattles are brittle and can easily break off. You **cannot age** these snakes by counting the number of rattles on the tail.

Rattlesnakes live in a wide variety of habitats, from sea level to nearly 11,000 feet. The desert southwest has the most diversity, and Arizona has the largest number of species, at 13.

Mojave Rattlesnake *Crotalus scutulatus*

featured creature
THE WESTERN DIAMONDBACK RATTLESNAKE
Crotalus atrox

DESCRIPTION: This heavy-bodied snake is typically brown to gray in color, with diamond-shaped dark blotches lining its back. Above the rattles, very distinctive black and white bands encircle the tail. It can grow to be over 6 feet long but is usually smaller. Two dark lines on each side of their triangular-shaped head run from their eyes to their jaws. These snakes can be aggressive and will hold their ground when threatened. Only its close relative, the Eastern Diamondback, gets bigger and is more dangerous.

RANGE: The Western Diamondback has a large geographic range in the southwestern states. It is found in Texas, New Mexico, Arizona, Oklahoma, California, and the northern half of Mexico. A small population exists in the very southern tip of Nevada.

HABITAT: Herpetologists call the Western Diamondback a "generalist" when describing its habitat. It lives in arid deserts, forests, grassy plains, rocky hillsides, and coastal flats. They are found at elevations from sea level to 7,000 feet.

DIET: These nocturnal ambush predators sit and wait for prey to pass close by. The heat-sensing pits on their top jaw can detect any warm-blooded animal that comes near. They feed on all small mammals within their territory, including mice, rats, squirrels, gophers, and rabbits. A good meal may sustain them for several weeks.

REPRODUCTION: Western Diamondbacks often travel to a community den to brumate (hibernate) in winter. A den may contain 100 or more snakes. In spring, they emerge, and males will "wrestle" one another to gain breeding rites. After mating, the female will give birth to living young in late summer. The young rattlers may stay close to the female for a few days but then are on their own, fully loaded with venom.

GUS' PAGE

Holding a pet python
(I was not allowed to hold a cobra)

THE KING COBRA

Ophiophagus hannah

DESCRIPTION: The longest venomous snake in the world is the king cobra. They are long and slender and can grow up to 18 feet. Their color can be dark brown, olive green, and almost black or yellowish. The king cobra has long vertebrae in their neck, that can expand to form a hood. This cobra has large amounts of venom, enough to kill an elephant or 20 humans. The average size of an adult is 10-12 feet.

RANGE: They live in northern India, southern China, throughout the Malay Peninsula, western Indonesia and the Philippines. In some countries, like Vietnam, they are protected.

DIET: King cobras eat mostly other snakes, venomous or not. They sometimes eat lizards, frogs, small mammals, and birds. The genus name, Ophiophagus, means snake eater.

HABITAT: They live in rainforests and bamboo thickets in tropical areas. They may also live in mangrove swamps and high-altitude grasslands. They hide underground or under rocks and trees. They are not normally aggressive but will raise their body a few feet off the ground and display their hood when they feel threatened.

REPRODUCTION: Male king cobras will often stay near the same female for several years during the breeding season. Females will make a nest of fallen leaves and rotting vegetation. They are one of the very few snakes that will construct a nest. The female will guard the clutch of eggs (up to 50) for two months or more. The babies are on their own when they hatch.

FUN FACTS and EXCEPTIONS SNAKES

Banded Sea Krait
Laticauda colubrina

Sea snakes are found in the warm coastal waters of the Pacific and Indian Oceans. Although highly venomous, they are not aggressive and pose almost no threat to humans. Related to cobras, their venom is a neurotoxin that works to paralyze the fish and eels that make up the bulk of their diet.

Some sea snakes can grow to be 5 feet long. Some can stay submerged for as long as 8 hours as they absorb oxygen through their skin. A few sea snakes give birth to live young in the water, and their young are usually quite large.

About 7,000 cases of venomous snake bites occur each year in the United States. With anti-venom drugs available, few people in North America ever die. But even if a person lives, the bite can be extremely painful, dangerous, and very expensive to treat. Around the world, anti-venom is often not available, and more than 80,000 people die from snake bites each year. Most deadly snake envenomations occur in India, which averages 11,000 deaths per year. These death statistics may be much higher as many people do not receive medical attention and die at home.

Inland Taipan
Oxyuranus microlepidotus

Probably the most venomous of all snakes is the Inland Taipan of central east Australia. It is said that the venom in one bite from the taipan is enough to kill 100 people. The inland taipan is shy, uncommon, and mild-tempered, so they are not considered to be the most dangerous of venomous snakes.

Snakes, like other reptiles, are **ectotherms**, meaning that their body temperature and activity level are dependent on the surrounding environmental conditions. If it gets too cold, they **brumate**, a kind of hibernation taking place in warm shelters. If too hot and dry, they may **aestivate**, a process of finding a cool shelter where they can avoid water loss and lack of food.

• Around 30% of snakes give birth to living young. All others lay eggs.
• Depending on their size, snakes may have from 600-1800 bones in their body. Humans have 206.
• Only 20% of snakes are venomous. Of these, less than 10% are considered deadly to humans.
• Some animals like opossums, the honey badger, and mongooses are immune to snake venom.
• Snakes shed their skin as they grow. Under ideal feeding conditions, they may shed 3-6 times a year. With flexible jaws, a snake may eat an object much larger than its head.

Snake Poems

Momma Snakes' Advice

A rattlesnake moved into town
He thought he'd have a look around His
mother gave him good advice
"In town you'll find more rats and mice"

"But of the people be aware
They will not like your empty stare
Your venom and your rattled tail
May cause you grief but do not fail"

"Make sure you hide and you'll be fine
Enjoy the rats on which you dine
And know that you provide a service
For folks hate rats they make them nervous"

"The day may come when you're respected
But till then go undetected
And when you clean out all the pests
Come back home and have a rest"

"Then find a mate and procreate
You'll have time its not too late
Then send your offspring back to town
And have them take a look around"

BOA OR PYTHON?
Turtles lay eggs and most lizards do
But snakes may have a trick or two
While boas give birth to living young
Pythons have shells surrounding 'em
So what's the difference tween the two?
NOTHING when they're constricting you!

Rock python constricting monkey

Extracting ratttlesnake venom

Red Diamond Rattlesnake
Crotalus ruber

! Snake Bite Advice !
If bitten by a rattlesnake
Here's all you need to know
Throw away the snake bite kit
Forget about the tourniquet
And to the ER go!!!

CROCODILES AND ALLIGATORS

There are 24 Species of Crocodilians found world wide
Order: Crocodylia

Cuban Crocodile *Crocodylus rhombifer*

American Alligator *Alligator mississippiensis*

Spectacled Caiman *Caiman crocodilus*

Gavial *Gavialis gangeticus*

Some crocodile specialists claim there are 23 crocodilians and others say there are actually 28 species. Either way, crocodilians make up only **one-quarter of one percent** (0.0025) of all the world's reptiles.

Saltwater Crocodile *Crocodylus porosus*

CROCODILIANS

The 24 species of crocodilians are found in 90 countries around the world. As a general term, crocodilians include crocodiles, alligators, caiman and gavial. The three families of crocodilians are:

Alligators: The American Alligator, Chinese Alligator, and the Caimans that are found in Mexico, Central, and South America.

Crocodiles: The family of true crocodiles that are found in both the New and Old World. Average adult size ranges from 5-18+ feet.

Gavial: The Gavial and Tomistoma are two very large animals with long, narrow snouts. Gavials live in India and Nepal. Tomistomas are found in Malaysia and Indonesia. Both are endangered.

Crocodilians are large reptiles with four short legs, thick-plated skin, and long, powerful tails. Their eyes, ears, and nostrils are high on top of their head, allowing them to get close to their prey in shallow water while their body remains submerged. All species of these semi-aquatic reptiles, crocodiles, alligators, caiman, and the gavial, have similar streamlined bodies and anywhere from 80 to 100 conical teeth.

Female Croc Taking Hatchling to Water

In general, crocodilians are tropical and subtropical reptiles that are aquatic but come on shore to lay their eggs in nests built by the female. They live in lakes, rivers, marshes, and estuaries. While most are found in freshwater, some can tolerate salt water.

Crocodilians are carnivores (meat eaters). Young crocodilians may eat insects at first, but all of these reptiles will eventually eat fish, crustaceans, birds, other reptiles, and mammals. Crocodilians don't chew their food and will crush their prey with extremely powerful jaws, and rip it apart to swallow whole.

Nile Crocodile Hatchlings

Most reptiles show very little parental care once their eggs have been laid, but crocodilians are different. Female crocodilians will protect their nests from predators once the eggs are buried.

When the young are ready to hatch, they grunt, which attracts her attention and she will dig open up the nest so the young can escape. Huge mother crocodiles are also known to gently gather the hatchlings in their mouths and carry them to the water's edge.

CROCODILIANS
Some examples

THE SALTWATER CROCODILE
Crocodylus porosus

The Saltwater crocodile is the **largest reptile in the world.** Males average 17 feet in length, and can easily weigh 1000 pounds. It has the largest range of all crocodilians. Today they are found on the eastern coast of India, through most of SE Asia and northern Australia.

The "salty's" hide is considered to be the **finest leather of all crocodilians.** They were over-hunted in Australia but are now protected with populations estimated to be near 100,000.

Saltwater crocodiles hunt close to the shore in fresh, salt and brackish waters. The will eat most anything they can catch, and have been known to eat sharks.

On occasion, a salty may venture out to sea and travel long distances. They show up on remote islands where they prey on nesting sea turtles and their young.

Saltwater crocodiles have the **strongest bite force** of all animals at 3700 psi (pounds per square inch). In humans, the bite force is around 150 psi.

THE BLACK CAIMAN
Melanosuchus niger

Caimans are members of the Alligator family and live in Mexico, Central and South America. Most of the six species of caiman are small-averaging 5-7 feet.

The Black caiman is a monster. They can grow to 14 feet. They are bigger than the American alligator, on average. Black caimans live in the Amazon Basin in eight countries of South America. They prefer shallow, fresh water habitats.

They were hunted to near extinction by the 1970's. With conservation efforts they have made a comeback. However, poachers still kill the Black caiman for its beautiful dark hide and tasty meat.

Young caimans eat insects, small fish, and frogs. Adult Black caimans feed on catfish and piranhas, but are also known to eat turtles, and large animals like deer, otters and capybaras.

Small caimans are eaten by a host of predators. Caiman up to 5 or 6 feet have been captured and eaten by Jaguars.

featured creature
THE AMERICAN ALLIGATOR
Alligator mississippiensis

DESCRIPTION: The American alligator is a very large reptile. Adult males are larger than females, and on average are 10-11 feet long and weigh 300 pounds or more. Their jaws are broad and rounded. Their limbs are short and thick with webbed feet adapted for swimming. Their bodies are armored with thick bony plates called osteoderms. The alligator's tail accounts for half its body length. The muscular tail helps the alligator swim fast—twice as fast as any human.

Gators are olive black in color as adults. Young are black with yellow crossbars. They have 80 teeth which are replaced throughout their lifetime as they wear down or are broken off. The position of their eyes, ears and nostrils is on top of their head allowing them to submerge their entire bodies and still see, hear and breath as they ambush prey at the water's edge.

RANGE: Alligators are found throughout Florida and Louisiana and in coastal wetlands of Georgia, North and South Carolina. They are also found in Southern Mississippi, Alabama, Oklahoma, Texas and the southern tip of Arkansas. Once threatened with extinction there are now **one million alligators** living in Florida alone.

DIET: Young alligators feed on insects, small fish, snails, crayfish and frogs. Adults eat turtles, snakes, fish, mammals and birds. They swallow their prey whole.

HABITAT: Alligators live in a variety of watery habitats. They prefer fresh water lakes, ponds, marshes, swamps and slow moving rivers. When water levels are low alligators often congregate in large numbers.

A pelican is an uncommon meal for a gator

Alligators gather as water levels drop

REPRODUCTION: Male alligators fight for the rite to breed with females in the spring. Females will build a nest of sticks, grasses, leaves and mud on dry land and lay between 25-50 eggs. A female will protect the nest until they hatch, and the young for a short time.

featured creature
THE AMERICAN ALLIGATOR

A VERY VOCAL BEAST: Baby alligators make high-pitched grunts when threatened, which attracts the attention of a protective mother. Adult gators hiss loudly when disturbed. A male alligator BELLOWS during the breeding season to show its superiority. He will raise his head, elevate his tail, and emit a loud roar. The water over his back vibrates and makes the water dance. It is a very impressive display, especially when several males join the chorus. Females also bellow, but not as much.

Male Alligator "Bellowing"

A CONSERVATION SUCCESS STORY: In 1962, commercial hunting of alligators was banned. In 1967, alligators were designated as an **endangered species.** With state and federal protection the populations rebounded dramatically. They had come close to extinction, but now, an estimated five million gators live in the southeastern United States.

A KEYNOTE SPECIES: The value of the alligator in the environment cannot be overstated. When seasonal ponds and swamps dry, they dig deep caves where water collects, which provides a source of life for other creatures. Aquatic insects, fish, turtles, water snakes, frogs, and other species take refuge in these gator holes. Mammals and birds find a place to drink when water is scarce. **The alligator is a life source for all.**

Gator on bank

DANGER TO US: Alligators only become **dangerous when uneducated people feed them.** They have been known to grab domestic dogs off the bank of a pond and to attack a human swimmer in a lake. **DO NOT SWIM** anywhere where alligators are known to be. Do not walk pets close to the water's edge. Don't feed or harass them. Just observe them, and you will be safe.

GUS' PAGE

THE GAVIAL
Gavialis gangeticus

DESCRIPTION: The gavial is a very large crocodilian with a long and narrow snout. They are also known as gharials. Their orange-green hide is very smooth, and valued for leather products. Their leg muscles are not as strong as other types of crocs, and they have a hard time walking on land, but with webbed feet, they are excellent swimmers. They may grow to be 12-15 feet long and weigh 500 pounds. Gavials, which were over 19 feet long and weighed one ton, used to exist.

RANGE: Gavials are found in northern India and Nepal, but in only 3% of their original range. They are listed as CRITICALLY ENDANGERED. Mining, agriculture, over-fishing, and pollution have destroyed their habitat.

DIET: Fish make up most of the diet as their long, narrow snout can move quickly through the water. Young gavials and adults will also eat frogs, crabs, and insects.

HABITAT: Clean fast flowing streams and rivers are what they prefer. In large rivers, they like to bask on sandbars. Most rivers where they are found are so polluted that they can no longer live in them.

REPRODUCTION: As they mature, the male gavial gets a large knot on the end of his snout. The knot, called a ghara, helps them to make a loud hissing sound to attract females. The females lay about 30-60 eggs in a dry sand nest. Their eggs are the largest of all crocodilians.

FUN FACTS and EXCEPTIONS CROCODILIANS

Crocodilians are the only reptiles with a four-chambered heart. This advantage allows the animal to lower its blood flow and preserve oxygen when submerged in water. **There are no reproductive exceptions, as all crocodilians lay eggs** in nests made of vegetation or dug in the sand. What is exceptional is that females protect their nests and help the hatchlings escape.

All crocodilians vocalize in one way or another. Some species make sounds underwater that can travel far. These noisy reptiles use the sounds to declare their territory and attract mates.

Crocodile hides are a valuable source of durable and beautiful leather. All 24 species have suffered from the effects of hide hunters since the 1800s.

Crocodile Leather Products

Commercial crocodile farms have relieved some of the pressure on wild populations. Each year, around **1.5 million hides** are produced in 30 different countries. The meat is also sold but has much less value.

Crocs and gators will swallow small stones, called gastroliths, that help to grind food in the stomach. The strong acid in their stomach can dissolve and digest bones, horns and shells.

A saltwater crocodile named Lolong died in the Philippines in 2013. He was 20.2 feet long and weighed 2300 pounds. He had been wild and was only in captivity for two years.

A commercial crocodile farm

Crocodilians are usually considered to be solitary ambush predators, but some herpetologists have witnessed **cooperative hunting**. Crocs may join together in slow-moving waters, with one encircling fish while others charge in to feed on the concentrations. When large prey animals are caught, a croc will perform a **death roll** that rips the victim apart so they can swallow the pieces more easily.

In spite of attempts to protect all crocodilians, some are seriously threatened and on the verge of extinction. The list of **CRITICALLY ENDANGERED** crocodilians includes the **Orinoco, Philippine, Cuban, Siamese, African slender-snouted crocodile**, the **Gavial**, and the **Chinese alligator**.

CROCODILE POEMS

Gator or Croc?

Have you or have you not
Been taught the difference
'tween a **gator** and **croc**?

They both are huge reptiles
of this there's no doubt
and one difference shows up
in the shape of the snout.

The gator's top jaw
overlaps the bottom
and the teeth underneath
don't show that they got-'em

But please be aware
that they have quite a few
it's just that those teeth
remain out of view.

More slender the jaw of the big crocodile
and many will say that a croc seems to smile.
Teeth on the bottom stick out on each side
large and impressive with no place to hide.

Gators are common and not easy to scare
while crocs are elusive, shy and more rare.
So, if you see a croc and think it's a gator
Remember this rhyme and we'll "see ya later"

Crocodile showing teeth on upper and lower jaws

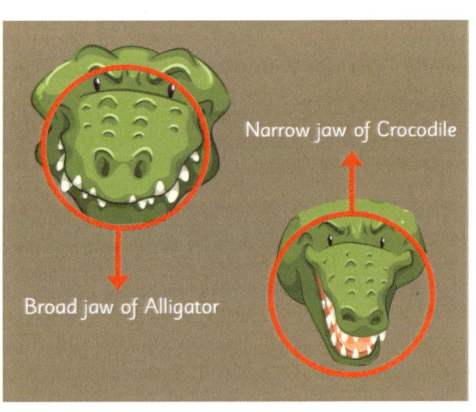

Apex Predator

Along the river and close to the bank
this apex predator deserves his high rank.

He doesn't care about what ventures near,
a turtle, large rodent or maybe a deer.

He glides near his prey, so cleaver and stealthy
sharp teeth and strong jaws-no wonder he's healthy

He'll strike with surprise, precision and style
a half ton of fury—the big crocodile.

He may not eat often-a good meal or two
STAY BACK FROM THE BANK-so he won't eat you.

Like him or not that's really okay
after millions of years he's not going away

Alligator teeth on lower jaw are hidden

32

LIZARDS

There are over **6,000** Species of Lizards
Order: Squamata

Green Iguana *Iguana iguana*

Komodo Dragon *Varanus komodoensis*

Gila Monster *Heloderma suspectum*

Leaf-tailed lizard *Uroplatus henkeli*

Eastern Collared Lizard *Crotaphytus collaris*

60% of all reptile species are lizards
25% of all lizards are Geckos
Australia has more lizard species than any other country

LIZARDS

Lizards have a dry, scaly skin, elongated bodies, long tails, four legs, eyelids, and ear openings, and they lay eggs-**usually**. However, there are legless lizards, short-tailed lizards, earless lizards, lizards without eyelids (geckos), lizards that give birth to live young, and some whose populations are all female. Lizards are found on all continents except Antarctica and exist in a wide variety of habitats from sea level to 16,000 feet.

Long-nosed Leopard Lizard *Gambelia wislizenii* eating Side-blotched lizard *Uta stansburiana*

A DIET OF GREAT VARIETY: Lizards, like iguanas, that eat plant material are called **herbivores**. Lizards that eat other animals are known as **carnivores**. Those that eat both plants and animals are **omnivores**. The small horned lizards feed almost exclusively on ants, while the Komodo dragon may eat an animal as large as a deer or a water buffalo.

SMALL TO EXTRA LARGE: The smallest lizard is a gecko that measures less than one inch long. The largest is the Komodo dragon, which may reach a length approaching ten feet. Most lizards encountered in the wild are within a range of 6 to 15 inches. The largest native lizard in the United States is the Gila monster, which is less than two feet long.

Green Basilisk *Basiliscus plumifrons*

Yellow-backed Spiny Lizard
Sceloporus uniformis

FROM DRAB TO A RAINBOW OF COLORS: Most lizards are rather drab in color and blend in with their environment, but during the breeding season, some may display colors that are truly beautiful. The loose skin under the chin (the **dewlap**) of male lizards can be brilliant when displayed as they try to attract a mate or ward off rival males. Chameleons are the most colorful of all lizards.

LIZARDS
Some examples

Green Iguana *Iguana iguana*

THE GREEN IGUANA: Male Green iguanas may reach a length of six feet and weigh up to 12 pounds. They are native to Mexico, Central and South America. Iguanas eat leaves, flowers, grass, fruit, and occasionally bird eggs and carrion (dead animals). Iguanas are both **arboreal** (tree dwellers) and **terrestrial** (ground dwellers).

During the breeding season, males turn bright orange. Like many other lizards, iguanas can regrow their tails. In their native countries, they are eaten and considered a good source of protein. They are often referred to as "Chicken of the Trees."

CHAMELEONS: There are more than 180 species of chameleons. Most are from the island of Madagascar, but they are also found in Africa, Spain, Portugal and Asia.

Chameleons have large bulging eyes that can rotate separately, giving this lizard a 360-degree view. Their slender, sticky tongues may be twice as long as their bodies, helping them to "reach out" and catch the insects they eat. Scientists say that their **amazing ability to change colors** is a response to temperature, humidity, light, and their current emotional state. They have prehensile tails and club like feet that make them excellent climbers. The smallest chameleon can fit on the head of a kitchen match, and the largest (Parson's chameleon) can reach a length of two feet.

Veiled Chameleon *Chamaeleo calyptratus*

Regal Horned Lizard *Phrynosoma solare*

HORNED LIZARDS: There are 14 species of horned lizards that are found in the western United States and Mexico. Most inhabit hot, dry deserts but a few survive at high altitudes, to 9,000 feet. They are hard to see as they blend perfectly into their surroundings.

Horned lizards are often referred to as "horny toads" because of their low profile and chunky bodies. They are generally small (3-6 inches) but have a **fearsome dragon-like appearance** with sharp horns on their heads above their eyes. Snakes and other predators find them hard to swallow. Ants make up the majority of their diet. Some species of horned lizards can squirt blood from the corners of their eyes into the face of a would-be predator. It is foul-smelling and a good deterrent.

featured creature
THE GILA MONSTER
Heloderma suspectum

DESCRIPTION: Gilas are the **largest native lizards in the United States.** They have an elongated body, fat tails, a short neck, and a large flat head. The average size of an adult monster is between 14 to 16 inches. They are beautifully colored with an orange, pink, or yellow back partially covered with blotches or bands of black. Their skin is tough and beaded. Under each small bead is a tiny bone called an **osteoderm**, which provides an armor rarely found in other reptiles. The genus name *Heloderma* actually means studded skin. They have a black forked tongue. **Gila monsters are the only venomous lizard found in the U.S.** Venom in lizards is rare.

Gila and Tortoise

HABITAT: Gila monsters live in deserts and semi-desert areas. They may inhabit grasslands and scrublands but prefer rocky foothills, where they use rock shelters to avoid extreme heat and cold. They will venture into open desert areas when searching for food and when the weather is mild. They may share a shelter with a tortoise as long as the tortoise does not have a clutch of eggs nearby.

RANGE: Gilas are found in SW Utah, southern Nevada, extreme SE California, SW New Mexico, western and southern Arizona, and the northern part of the state of Sonora, Mexico. They are native to the deserts of the Mojave, Sonoran, and Chihuahuan. The Sonoran Desert in Arizona, with more abundant rainfall, is where Gila monsters are most commonly seen.

Eating a quail egg

DIET: These fascinating lizards are nest-feeding specialists. They eat the eggs of birds, snakes, lizards, and desert tortoises. Gilas have sharp claws and strong feet designed for digging. They swallow quail eggs whole and eat the young of desert rodents and rabbits. Three to five good meals can sustain a Gila monster for an entire year. They store fat in their tails.

Photo: Cameron Rognan

featured creature
THE GILA MONSTER
Heloderma suspectum

SHELTER FIDELITY: Gila monsters may use the same overwintering shelter and summer shelter for years. This is called "Shelter Fidelity." When they are actively foraging in the open desert, they tend to use familiar shelters for days at a time. Other Gilas within their range will use the same shelter that one lizard has moved from. These usually solitary lizards have home ranges of 50 acres or more. Several females may be found within the home range of a single male. A monster may stay in an overwintering shelter for four or five months.

MALE COMBAT: Male monsters will protect their territory in the spring mating season. A male will fight for hours with a competitor and try to pin it to the ground and chase it away. Bites occur, but normally, the loser walks off unharmed to wait for another chance. Breeding happens in shelters and is rarely witnessed. Hatchling Gilas are mainly nocturnal and are not often seen.

VENOM FOR DEFENSE: Unlike rattlesnakes that use their venom to kill their prey, Gila venom is used in defense against predators. When they bite, they do not like to let go, and the venom from their lower jaws seeps quickly into the attacking animal, causing agony and pain. A bite to a human will not cause death, but it is a serious medical ordeal, and a visit to the hospital is recommended. **NEVER TOUCH OR TRY TO CATCH A GILA MONSTER, AND YOU WILL BE SAFE.** They are slow-moving and not aggressive.

LIFE OUT OF SIGHT: Herpetologists estimate that Gila monsters spend 80 to 90 percent of their time hidden in shelters and out of sight. They are diurnal when the weather is mild and can be nocturnal when the heat of the day is detrimental to their survival. They seem to be most active when temperatures are between 75 to 90 degrees Fahrenheit. When surface-active, some say that while walking, it looks as though their front end does not know where their back end is going. It is a comical gait.

A SNAKE'S TONGUE: Like snakes, Gila monsters have a tongue that is forked at the tip. A flicking tongue can detect chemicals in the air and on the ground that will lead a Gila to food, a mate, or a familiar shelter. The forked tongue of Gilas, and of monitor lizards, provides a definite advantage over the tongues of most lizards which use theirs to secure prey and bring it to the mouth. Some scientists state that the Gila is more closely related to snakes than to other lizards.

GUS' PAGE

THE KOMODO DRAGON

Varanus komodoensis

DESCRIPTION: Komodo dragons can grow close to ten feet long and weigh 300 pounds. Most are shorter and lighter. They have flat heads, huge, strong tails and sharp claws. They have powerful jaws, sharp teeth and a forked tongue that can sense in any direction. They are the world's largest lizard.

RANGE: Komodo dragons only live on five islands in Indonesia. They mostly live on the islands of Komodo and Flores. In 1980, a National Park was built for their protection.

HABITAT: Komodo dragons can live in savannas or grasslands. They can also live in rainforests. They sleep in burrows at night to keep warm.

DIET: Komodo dragons are meat eaters. They eat pigs, deer, water buffalo, and smaller dragons. They swallow their food whole and can eat 80% of their body weight in one feeding. The venom and bacteria in their mouth help to kill their prey.

REPRODUCTION: Female Komodo dragons can lay up to 30 eggs that are the size of a large orange. They will protect their nest until the eggs hatch. Females can sometimes lay fertile eggs without having to breed with a male dragon. In some zoos that separate males and females this has happened.

Gus watching a Gila monster seeking shelter

Gila monsters and Komodo dragons are members of the same superfamily, Varanoidea. Both have forked tongues and venom glands.

FUN FACTS and EXCEPTIONS LIZARDS

There are only a few hundred species of legless lizards found throughout the world.

Lots of smaller lizards are capable of regenerating a lost tail, although it will never look quite the same as the original. Tail loss, called caudal autumy, can help a lizard evade a predator. When the tail breaks off, it wiggles for a while, causing confusion and allowing the lizard to escape.

Many small lizards run fast; however, the large, bulky, Black spiny-tailed iguana, Ctenosaura similis, has been clocked at speeds of 21 miles per hour.

Eastern Glass Lizard
Ophisaurus ventralis

Frilled Lizard
Chlamydosaurus kingii

Most lizards lay eggs and do not protect the nest. There are some that are **viviparous** (giving birth to living young). Several species of skinks are an example.

A few species of female western whiptail lizards and some others give birth to young WITHOUT having to breed with a male. This phenomenon is known as **parthenogenesis.** The offspring, in this situation, will normally be all female.

Most all lizards are land dwellers (terrestrial). Basilisk lizards from Central and South America can run on water for a short distance. The Galapagos marine iguana, Amblyrhynchus cristatus, feeds on algae in the sea and **can stay submerged for an hour.**

Marine Iguana *Amblyrhynchus cristatus*

Clouded Monitor *Varanus nebulosus*

About 80 species of Monitor lizards are found in Africa, Asia, and Oceania. In Australia, there are 27 species. These large lizards of the genus Varanus are often killed for their meat and beautiful hides.

Lizard Poems

Eyelash Viper eating lizard

Lizard Anatomy Lesson

Most lizards have legs but in snakes they are missing.
Lizards have ears and can hear a snake hissing.
With eyelids a lizard can blink for a minute.
Blink a minute too long and the snake has it in it!!!

Many Crazy Lizards

When you think about a lizard
What first comes to your mind?
Go anywhere to see them
They are not that hard to find.

Horned lizards small and cryptic
This nobody denies
If close to being captured
They squirt blood from near their eyes.

The Komodo is the largest
it can kill and eat a deer
With sharp teeth and potent venom
They're a lizard we should fear.

Geckos are so cute
As they climb up on the walls
Most are very vocal
Making loud, annoying calls.

Chameleons look so funny
They have large and bulging eyes
With long and sticky tongues
They catch insects: gnats and flies

Lizards are amazing
Large monitor, small skink
They look like little dinosaurs
A pre-historic link.

Small gecko falls prey to
Black widow spider
Photo: Cameron Rognan

Thorny Devil
Moloch horridus

TURTLES

There are approximately 356 Species of Turtles
Order: Testudines

Pancake Tortoise *Malacochersus tornieri*

Ploughshare Tortoise *Astrochelys yniphora*

Desert Tortoise *Gopherus agassizii*

Galapagos Tortoise *Chelonoidis nigra*

The Ploughshare and Pancake Tortoises are listed as Critically Endangered.
The Mojave Desert Tortoise and the Galapagos Tortoise are Threatened species and on the verge of Endangered.
About 42% of all turtle and tortoise species are listed as Threatened.

Spiny Softshell *Apalone spinifera*

TURTLES

Turtles are easy to recognize. They have four legs and a tail. What makes them so different is the shell, which is a modification of the rib cage and spine. The upper shell is called the **carapace**, and the lower shell is the **plastron**. In most turtles, the shell is hard, strong, and built for protection. Some turtles have a soft, more skin-like shell.

The word "**TURTLE**" is a catch-all name used to describe these reptiles as a group. To be a little more specific, some biologists will often use the following terms to describe the environment in which they live.

TORTOISES - live on land and usually have high-domed shells.
TERRAPINS - live on land and in brackish water around swamps, estuaries, and bays.
TURTLES - live in water (fresh or salt), have streamlined shells and webbed feet or flippers.

Turtles are found on all continents except Antarctica. More species of turtles live in southern Asia, and in the southeastern part of the United States than anywhere else.

All turtles lay eggs on dry land, so even a huge female sea turtle must drag herself up a beach, above the tide line, to dig a nest chamber. Turtles inhabit about every type of environment there is, including hot deserts, grasslands, forests, lakes, swamps, rivers, tropical jungles, and, of course, the oceans.

Turtles can be **carnivores** (meat eaters), **herbivores** (plant eaters), or **omnivores** (eating both meat and vegetation). Turtles are mainly **diurnal** reptiles (active during the day).

Star Tortoise
Geochelone elegans

Because most turtles are mild-mannered and some stunningly beautiful, they are collected for the pet trade from around the world. Illegal collection has pushed the existence of some species almost to the brink of extinction. Declining numbers can also be attributed to habitat destruction, pollution, and consumption of turtle meat and eggs.

Leatherback Sea Turtle
Dermochelys coriacea

The majority of turtles are medium in size and may only weigh a few pounds.
A Leatherback male sea turtle can weigh well over 1000 pounds. The giant tortoises of the Galapagos and Aldabra Islands may reach 500 pounds or more. One of the smallest species of turtle is the Padaloper tortoise, Chersobius signatus, of South Africa. It weighs in at a mere 5 ounces and is less than 4 inches long as an adult.

TURTLES
Some examples

SEA TURTLES: There are 7 species of sea turtles found in oceans around the world. They nest on tropical and subtropical beaches. The two smallest sea turtles are the Kemp's and Olive Ridley, which may reach 100 pounds. A massive male Leatherback, by contrast, may tip the scales at 1000 to 1500 pounds.

Hawksbill *Eretmochelys imbricata*

Sea turtle diets are quite varied. Leatherbacks feed on jellyfish. Green turtles eat seagrasses. The Hawksbill turtle consumes sponges, which left unchecked, can severely damage a coral reef. Other species of sea turtles eat fish, conchs, bivalves, shrimp, jellyfish, sponges and seaweed. In other words, sea turtles help keep the oceans balanced and healthy.

Sea turtles may travel thousands of miles each year to their favorite feeding grounds but usually come back to the same beach where they were hatched to lay eggs. Most species are solitary nesters, but the Kemp's and the Olive Ridley are known to nest in a mass. This strange occurrence happens when hundreds or thousands of females gather near the beach and come ashore in what is called an **arribada**. As many as 300,000, or more, will nest over a period of several days. They will repeat this amazing act a few times at two-week intervals. Costa Rica is well known for arribadas of Olive Ridleys. In Mexico, the rarest of all sea turtles, the Kemp's, is a mass nester at Rancho Nuevo.

GIANT TORTOISES: Giant tortoises are found on the Aldabra Atoll in the Indian Ocean and on the Galapagos Islands off the coast of Ecuador. Twelve of the original 14 different-looking Galapagos tortoises survive today on the larger islands in the chain. In 2012, a 100 year old male died in captivity. Lonesome George was the last male of his species from La Pinta Island. Giant tortoises are herbivores eating grasses, shrubs, flowers, cactus pads, and fruits.

There was once a very large population of tortoises in the Galapagos. Over a span of the last 200 years, whaling ships, pirates and fur traders raided and collected tortoises by the thousands with no concern for their actions.

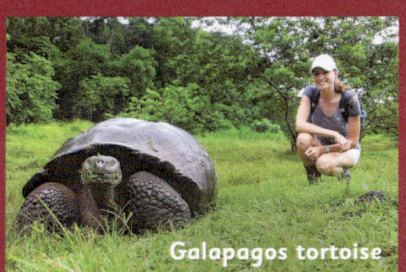

Galapagos tortoise

Tortoises are a great source of protein and the easily caught giants could be stored on a ship for nearly a year without food or water until eaten. It has been estimated that 200,000 tortoises were taken during that devastating period.

Giant tortoises now receive full government protection. However, introduced rats, pigs, goats and feral dogs on the islands are taking a toll on tortoise hatchlings. The population of all species on the islands is estimated to be a scant 20,000-25,000 tortoises.

featured creatures
DESERT TORTOISES

DESCRIPTION: The Desert tortoise reaches an adult size of between 9-15 inches. They have a high-domed shell, flattened front legs for digging, and elephant-shaped rear legs. Adults are brownish gray on top (the carapace) and yellowish tan on the bottom (the plastron). They are stout, with an average weight of 6-10 pounds.

RANGE: Desert Tortoises are found in eastern California, southern Nevada, southwestern Utah, western and southern Arizona, Texas and Sonora, Mexico. They inhabit the Mojave, Sonoran and Chihuahuan Deserts. There are now **five species** of desert tortoise recognized by herpetologists. In general, they look very similar, though minor differences in size and color are noticeable.

Eating Coyote scat with bird feathers

HABITAT: These tortoises live in very harsh environments. They inhabit semi-arid grasslands, gravelly washes, canyon bottoms, and steep rocky hillsides. Despite their low profile and short legs, they are extremely agile and can climb up boulder-strewn cliffs to incredible heights. Tortoises spend most of their lives hidden in burrows they dig for themselves under rocks, on wash banks, or under a bush in the open desert. They usually have many burrows they use throughout the year.

DIET: Tortoises are herbivores (plant eaters). Wildflowers, grasses, dry shrubs, and the pads, flowers, and fruits of cacti are all consumed. When available, they will drink water from any source they find and can store enough water in their bladders to last for a year.

REPRODUCTION: Male desert tortoises will fight one another for territory and the right to breed with females. With their elongated **gular horn** (the pointed extension on the lower shell), they attempt to flip their competitor on its back. Females can store sperm for a year or more if conditions are not suitable. Normally, a female will lay 4-8 eggs in early summer and may nest again a time or two if food and water are really available. Desert tortoises reach sexual maturity when 15-20 years old. A desert tortoise may live for 50 years or more

Spring snack

featured creature
THE MOJAVE DESERT TORTOISE
Gopherus agassizii

Found in the Mojave Desert in California, Nevada, Utah, and Arizona, this tortoise is listed as **a threatened species.** They are Federally protected, meaning that it is illegal to touch, catch, or harass them.

No other reptile has been monitored and studied more than Gopherus agassizii. Their habitat is shrinking as solar fields, wind fields, and large housing projects invade the desert they call home. With every new proposed project where tortoises exist, wildlife consulting companies are hired to survey the area and, if necessary, translocate tortoises to an acceptable habitat. Translocating tortoises is rarely successful. For a host of different reasons, many cannot adjust and die.

Scientists describe desert tortoises as **"ecosystem engineers"**. Their excavated burrows provide cover and safety for numerous desert creatures like insects, snakes, lizards, burrowing owls and small mammals. Tortoises dig many burrows in their home range.

Tortoises in the desert hibernate (**brumate**) when the weather is cold. They also avoid the hottest time of year by staying in shelters where the humidity is greater and the temperature is cooler (**aestivation**). Surface activity is limited by temperature, which is why they are often hard to find.

Many Mojave tortoises are kept in captivity legally if they were acquired before 1989.
Some states have adoption programs specifying that only one tortoise can be held captive. Breeding and releasing tortoises back into the wild is strictly prohibited. Adoption programs become necessary when tortoise keepers realize that their "pet" will outlive them and want to find them a new home.

Baby desert tortoises are only 2 inches long when they hatch. Most never survive to become adults as their small size and soft shells leave them vulnerable to a wide variety of predators such as foxes, coyotes, badgers, and birds of prey. Ravens are one of their worst enemies. Dogs, allowed to run off leash, also kill baby tortoises.

GUS' PAGE

THE ALLIGATOR SNAPPING TURTLE

Macrochelys temminckii

African spurred tortoise
(There are no Alligator Snapping Turtles where I live)

DESCRIPTION: Alligator snapping turtles are the largest freshwater turtles in North America. They have a big head and strong jaws with a hooked bill. They can grow to 30 inches long and weigh up to 150 pounds or more. These turtles have a worm-like appendage on their tongue, which they wiggle with their mouth open to attract their prey, like fish. They have three ridges that run along their top shell from head to tail.

RANGE: This aquatic turtle lives in the southeastern United States. They are found in northern Florida, southern Georgia, Alabama, eastern Texas, and north in a few states in the Mississippi River drainage. In most states, they are very rare. They have been hunted for years as a source of food.

HABITAT: Alligator snapping turtles live in freshwater areas. They can be found in large rivers, swamps, and lakes. They prefer deep water and can stay submerged for 50 minutes before coming to the surface to breathe.

DIET: Alligator snapping turtles eat fish, frogs, snakes, crayfish, and other turtles. They are active hunters, feeding day and night. They will also eat dead animals (carrion). They sometimes, but rarely, eat plants.

REPRODUCTION: They mate in spring. Females lay their eggs on land two months after mating, and babies hatch after 100 days or more. Alligator snapping turtles reach their breeding age after 10 years. Adult turtles can live to be 40-50 years old.

FUN FACTS and EXCEPTIONS TURTLE

There are said to be only 3 remaining specimens of the Yangtze giant softshell turtle. It may be too late to save this turtle. Adults have been known to weigh 350 pounds.
All around the world, efforts are being made by dedicated scientists to conserve land and sea turtles. Sadly, **145 of the 356 species of turtles are considered threatened or endangered.** Since the 1800s, ten different turtle species have gone extinct.

As with crocodilians, turtle reproduction is straightforward. After mating, the female digs a nest with its hind legs and lays eggs on land. Turtles do not protect their nests or hatchlings.

Redfoot Tortoise
Chelonoidis carbonaria

Small turtles reach maturity much faster and lay fewer eggs than their larger relatives. A mud turtle may be ready to breed after 3 or 4 years and will lay just a few eggs. Sea turtles do not become mature and ready to breed, for 15 years or more. They may lay 50-80 eggs.
In 2019, on the Galapagos Island of Fernandina, a female giant tortoise, a subspecies thought to be extinct for 113 years, was found. Hopefully, more will be discovered.

The sex of turtle hatchlings depends on the incubation temperature inside the nest. Cooler temperatures produce male turtles, and warmer temperatures produce females. A typical nest will have different temperatures at various depths, so a mixture of males and females usually results.
Turtles, in general, have good eyesight, and many see color. Some turtles have good long-term memories, and wood turtles, for example, are said to be better at navigating a man-made maze than a domestic rat.

Turtle farming for the meat of softshell and snapping turtles has been going on in Japan and China for over a hundred years. In the United States, farming of red-eared sliders for the pet trade is a big business. **The red-eared slider has become the most invasive turtle species** worldwide as they escape captivity and out-compete native turtles in natural environments. **Never release** pet turtles into lakes, rivers or ponds.

Turtle Farm- red-eared slider
Trachemys scripta elegans

Turtle Poems

A Sea Turtle's Devotion

Sea turtle success means a trip to the land
to lay eggs on a beach in the soft summer sand.
With paddle-like flippers somehow she is able
to dig out a nest that is deep, round and stable.

As eggs start to drop she falls into a trance
and nothing will stop her as this is her chance
to continue the cycle that's gone on forever.
An arduous task, an amazing endeavor.

Her work is not done till she covers and pounds
sand over the eggs, sand flinging around
in hope of disguising the nest she has made
from predators seeking the clutch she has laid.

Tired and weak she heads back to the ocean.
Hours of labor, exhaustive devotion.

In ten years or twenty upon the same beach
one of her offspring may finally reach
the same destination where she came to life
and somehow survived the dangers and strife.

Green Sea Turtle
Chelonia mynas

Eastern Box Turtle
Terrapene carolina carolina

A Fabulous Beast

They live in the oceans, they live on dry land
They are very resilient, they lay eggs in the sand.
For millions of years they've conquered the hurdles
A likeable lot, who doesn't like turtles?

Crocs are impressive and powerful beasts
Snakes are the ones most people like least
And lizards are fine as reptiles go
But they're not the ones that most of us know.

A turtle may be as small as your fist
Or as big as a car and hard to dismiss
Gentle yet tough and ever so fertile
A lovable beast—the fabulous turtle

48

SEXUAL DIMORPHISM

Sexual dimorphism is defined as the difference in form between individuals of different sexes in the same species. In other words, males and females of the same species may look quite different from one another.

In reptiles and amphibians, the difference may be in size, color, head and tail shape, eye color, or other body features. In many species, there is very little difference at all.

Sitana ponticeriana,
Fan-throated Lizard

Like everything in science, **there are no hard and fast rules** when talking about sexual dimorphism. For instance, a male sea turtle is much larger than the female. Most female freshwater turtles are larger than males.

Male Green Iguana
Iguana iguana

In snakes, the female **ANACONDA** dwarfs the smaller males, but in rattlesnakes, males are a little bigger. **Frog females** are almost always larger than the males. In many species of snakes, lizards and turtles, the male and female may be very similar in size, so it can be difficult when you see one in the wild to determine what sex it is.

A male Green iguana becomes bright orange during the breeding season, which helps him to get the attention of the females.

A female leopard lizard from the desert southwest has beautiful orange and red spots all over its head and body during the reproductive months, which are absent in the male. Yet, in many species of lizards, the males are more colorful during the breeding season.

So dimorphism can be a little confusing when you are just starting out to learn about reptiles and amphibians, but it is just one more fascinating aspect in the study of herpetology.

Long-nosed Leopard Lizard
Gambelia wislizenii

FLORIDA'S EXOTIC NIGHTMARE

Florida has a problem. Non-native reptiles and amphibians are everywhere in South Florida, where the climate is subtropical. A hard freeze is rare in the southern part of the state, and the exotics have thrived.

There are **142 native reptiles and amphibians** in the Sunshine State. The Florida Fish and Wildlife Conservation Commission (FWC) claims there to be **63 non-native established species.** By established, they mean that these animals are breeding and producing offspring over a number of years. There are now more non-native lizard species in Florida than there are native species. In fact, Florida has more non-native, invasive species than anyplace else on earth.

Some of these animals arrived on cargo ships, hidden in food or plant imports. Some have escaped from zoos, especially after major hurricanes.
However, the majority of animals are ones that have escaped or been released by uneducated and irresponsible pet owners.

The Cuban tree frog (Osteopilus septentrionalis), the Cane toad (Rhinella marina), the Burmese python (Python bivittatus), and the Green iguana (Iguana iguana) are the best-known, and the ones that cause a lot of problems.
The Cuban tree frog eats native tree frogs, lizards, small snakes, and insects. The mucus on the skin is irritating to humans. Found all throughout Florida, they have added to the decline of native frogs.

Burmese Python

The Cane toad, or marine toad, can grow to 9 inches, and eats native species of reptiles and amphibians, and will dine on pet food left outside.
Pets and other mammals eating poisonous toads get sick and may die.
The Green iguana is a plant eater, but they dig burrows under roadways, seawalls, and home foundations. Repair costs are enormous.

The Burmese python is the best-known of the exotics attacking Florida. Their numbers in the Everglades are staggering. They are big snakes and are believed to have caused small mammal populations to decline drastically.

ON A POSITIVE NOTE: The nesting of Loggerhead sea turtles on Florida's beaches has increased dramatically in recent years. In 1974, the author recorded only 52 turtle nests on a six-mile stretch of beach in Bonita Springs. In 2022, there were 238 nests recorded on a two-mile stretch of the same beach. Conservation efforts have made a difference.

FLORIDA'S EXOTIC NIGHTMARE

Nile Moniter *Varanus nilotic*

This five foot long lizard is fast and aggressive. In Cape Coral, Florida, for example, its population is reported to be over 1000 animals. They are semi-aquatic, swim well, and are also excellent climbers. **They eat anything they can catch,** including small pet cats and dogs. With sharp claws and teeth Nile monitors are extremely damaging to native wildlife. Their burrows along canals often damage sea walls in neighborhoods where they also frighten residents.

A native to Mexico and Central America, this large iguana grows to an average size of four feet. Though big and heavy, the spiny-tailed is considered to be the fastest lizard alive, running for short sprints at 21 mph. In Florida they are found up to Tampa on the west coast, and in Broward County to the Keys in the south. They are omnivores, but tend to be more herbivorous as they age.

Spiny black-tailed lizard *Ctenosaura similis*

Black and White Tegu Lizard *Salvator merianae*

Another large lizard that is rapidly spreading its range in Florida is the black and white tegu from Argentina. Growing close to five feet long, these omnivores are known to prey on the eggs of sea turtles, alligators, gopher tortoises, birds and small mammals. They survive the cold by living in deep burrows.

These colorful, small lizards (up to 24 inches) got established around Ft. Myers, Florida in 2002. From Yemen and Saudi Arabia, they seem to be able to withstand the cold winters even into south central Florida. Their impact on native wildlife may not be as dramatic as others, but they do compete with native lizards for food.

Veild Chameleon *Chameleon calyptratus*

REPTILE SIZES AND AGES

When it comes to recording the biggest snake or crocodile in the world **sizes are often greatly exaggerated.** Fake and manipulated photos occur frequently on the internet.

Captive bred reptiles in zoos and private collections may **grow larger and live longer** than animals in the wild, as they are fed on a regular basis and are not under the same amount of stress as those living in nature. There are records of captive snakes that have lived for over 35 years, for example. A Komodo dragon died at age 30 in a zoo in Calgary, Canada. **The information below is not meant to be exact.** There will always be bigger, heavier and older examples reported.

CROCODILES:
Saltwater Crocodile *(Crocodylus porosus)*
18-20 feet (5.5 meters)
2000 pounds (900 kgs)
65 year lifespan

SNAKES:
Anaconda *(Eunectes murinus)*
27 feet (8.2 meters)
400 pounds (181 kgs)
10-15 year lifespan

SEA TURTLE:
Leatherback *(Dermochelys coriacea)*
6.5 feet (2 meters)
2000 lbs (900 kgs)
30 year lifespan

TORTOISE:
Galapogos Tortoise *(Geochelone nigra)*
4.9 feet (1.5 meters)
880 pounds (400 kgs)
100 year lifespan

Author with Leatherback

LIZARD:
Komodo Dragon *(Varanus komodoensis)*
9 feet (2.7 meters)
300 pounds (135 kgs)
17 year lifespan

REPTILES AND AMPHIBIANS THAT WE EAT

Throughout the history of human existence, people have eaten reptiles and amphibians as they are a rich source of protein and low in saturated fats. Frogs, snakes, crocodilians, turtles, and lizards are all eaten around the world in one form or another.

FARMING: Reptile farming for meat and hides is becoming more widespread every year and everywhere. In southeast Asia, small landowners are finding that snakes and lizards are cheaper and easier to raise than cows, chickens, and pigs. They eat far less and are easy to house. Amphibian farming is more complicated, as the life cycle of frogs presents problems at each stage of development. Still, it is a huge business.

Frog legs

Bullfrogs are farm grown in large numbers in Brazil, Ecuador and Taiwan and shipped worldwide. They are native to North America. Unfortunately, some escape these foreign farms and devour native frogs and other small animals. Many Asian farmers are finding that some frog varieties do better than the bullfrog, which only eats live food. Their native frogs will accept manufactured food in pellet form.

Alligator and crocodile farms have cropped up all over the subtropics and tropical regions of the world. These are generally, large-scale operations and expensive to operate. Crocodile farms have taken some of the pressure off of the reptiles being taken from the wilds for their valuable hides and meat.

Lizards are probably the least eaten reptiles. Green iguanas of the American tropics are now being farmed. They are herbivores and, therefore, inexpensive to feed. Large monitor lizards in Australia and Africa are frequently captured for their meat and hides.

Snakes are raised for their meat, hides, and medicinal properties. In the small Chinese village of Zisiqiao, over 100 families raise 3

Crocodile on a grill, SE Asia

million cobras and pythons a year for domestic and world markets. Wild-caught snakes are often found in open-air markets all over Southeast Asia.

Turtles are a favorite food of many. Sea turtles that come ashore are vulnerable to human poaching. Eggs are dug up and sold, as some people believe there are health benefits derived from them. All sea turtle species are protected worldwide, but that does not stop some people from taking them.

Freshwater **snapping turtles** are a popular food item in the United States. The awesome and huge Alligator snapping turtle of the southeastern United States was almost "fished out" of the Mississippi River Basin before protection was awarded to the species.
In Thailand, a Chinese softshell turtle farm produces a million baby turtles each year. They grow fast and are sold to markets in East Asia.

The land-dwelling tortoises of the world have always been considered "fair game" for human consumption. Seafaring whaling ships that passed by the famous **Galapagos Islands** used to stop and send their crews on land to collect as many of the huge and amazing tortoises as they could find. They would store the tortoises below deck for months until they needed them for food. These giant reptiles are now protected, but their population is a tenth of what it once was.

REPTILES AND AMPHIBIANS AS PETS

Probably, most professional herpetologists grew up having a turtle, snake or frog as a pet. It might have been what got them interested in choosing their career. There is a lot to be learned by keeping a reptile or amphibian, but it is not a hobby for everyone. Care can be difficult and expensive. Terrariums and aquariums with special lighting, humidity, and temperature control fixtures have to be well maintained. Keeping the cage clean and the animal well-fed can be challenging.

The most popular reptiles in the pet trade are the Bearded Dragon and the Leopard Gecko. They are relatively easy to maintain. These two lizards are bred in captivity, but keep in mind that even today, a large percentage of reptiles are wild caught and sold illegally.

Bearded Dragon *Pogona barbata*

Reptile Shows and **Pet Stores** are good places to go to learn about an animal you may want and how to care for it. The internet may be better, as you can get many opinions and tips on a particular animal without being pressured to buy one.

Not a good pet choice

1. Some snakes, lizards, and turtles can live for a very long time. A snake may live for ten years, a lizard for 20, and a tortoise for 30-40.

2. Reptiles and amphibians don't bark or play with you like a pet dog would and, therefore, are often forgotten and sometimes neglected.

3. Snakes, like pythons and boas, can grow to extreme sizes and can bite hard and constrict even harder.

4. Keeping dead feeder rats or mice in the freezer may freak out your mom.

5. Getting rid of a pet reptile can be challenging, and releasing them into the wild **IS NEVER AN OPTION.** Florida, for example, has big environmental problems with Green iguanas and Burmese pythons that people have released in the Everglades.

But if you are truly interested in having a snake, turtle, lizard, or frog in your home and are committed to its care and well-being, then the rewards can be worth the effort. Perhaps you will be interested in becoming a professional herpetologist.

REPTILES AND AMPHIBIANS BY THE NUMBERS

The total number of reptile and amphibian species found on earth is and always will be confusing at best. This is partially due to the fact that scientists (taxonomists) often disagree about what animal is a true species or just a subspecies.

New discoveries happen occasionally, especially with amphibians like salamanders and frogs. The number of turtle and crocodilian species is pretty well known to science, and will probably not increase, and hopefully not decrease.

Below is an updated list of amphibian and reptile species known to science listed by some of the most trustworthy websites.

AMPHIBIANS;
- Frogs — 7,040 species
- Salamanders — 717 species
- Caecilians-eel like amphibians — 212 species

REPTILES;
- Lizards — 6512 species
- Snakes — 3709 species
- Turtles — 351 species
- Crocodiles — 24 species
- Amphisbaenians-Worm lizards* — 196 species
- Rhynechocephalia-Tuataras * — 1 species

*Amphisbaenians, or worm lizards, are odd reptiles, usually with no legs. They are rarely seen and are normally under 6 inches long. They live in burrows and feed primarily on insects.

*Tuataras are a rare and fascinating lizard-like reptile only found on a few islands around New Zealand. They live in a cool climate, grow to be 20 inches long, and are said to be the closest living animal on earth related to extinct dinosaurs.

White worm lizard *Amphisbaena alba*

Tuatara *Sphenodon punctatus*

NEW WORLD - OLD WORLD
The Continents

When reading history books or learning about plants and animals you will often see a reference to the **New and Old World.**

This is a simple, yet useful concept, you will encounter often when studying biology, literature, history and music. After the Americas were first discovered, historians started referring to them as the New World: **North, Central and South America**: the Western Hemisphere.

The Old World includes **Europe, Asia and Africa**. It is the Eastern Hemisphere.

Many reptiles and amphibians that may look and act alike are separated by thousands of miles of oceans between the hemispheres.
Pythons and boas are a good example but, **Pythons are Old World snakes. Boa Constrictors are New World reptiles.**

THE SEVEN CONTINENTS

However, this simplistic phrase has become complicated in today's world. Reptiles and amphibians have been transported across vast oceans for various reasons, and escaped animals from the pet trade has made the New and Old World designations for animals almost a joke.

Cane toads taken to Australia from the New World of South America to eat insects in the sugar fields, now eat everything they can catch, and are said to number 200 million. To make matters worse, anything that eats the toads gets sick or dies.

Pythons, monitor lizards and chameleons of the Old World wreak havoc on Floridas' environment, with little hope of eradication. The list of invasive species spread out all over the world in places where they don't belong is nearly endless.

FIELD GUIDES AND MEASUREMENTS

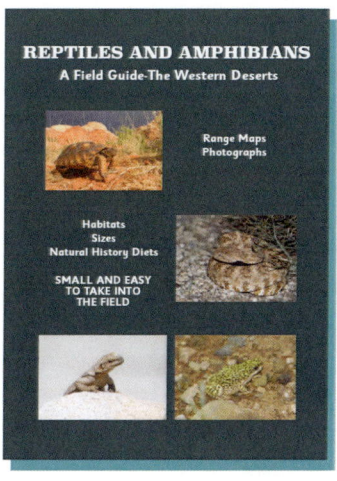

A good field guide is a valuable resource in the study of herpetology. Regional guides are best, as they are usually designed to be carried with you in a backpack or pocket. You may find one for your state or region that will make it much easier to pinpoint the reptiles and amphibians in your area.

A good field guide will include a range map for each species, and a description of the animals' normal habitat, along with photos or drawings. The size, color, diet and notes on its natural history should also be included.

The field guide will list the common and scientific name. Since common names vary from one area to another, the scientific name, (Genus and species), is important to know. Guides will provide size measurements in inches and feet in the United States and/or measurements in the metric system of millimeters and centimeters. This is important to know as the United States, and only two other countries in the world, still use the outdated system of measuring lengths and weights.

When measuring the size of a reptile or amphibian you will see the term **SVL** used.

This is the measurement from the SNOUT to the VENT, which is the cloacal opening at the base of the tail. Tail lengths, even of one species, may vary greatly so this SVL measurement is more accurate.

It is difficult to tell the sex of a reptile or amphibian by size alone. Male tortoise and sea turtles are bigger than females. The female anaconda dwarfs the smaller males. However, the majority of reptiles and amphibians are not dramatically different in size.

Range map of American alligator

After a day in the field, the internet is a great source of information. **Taking pictures from a distance**, without handling or disturbing the animal, is the best way to learn about their natural history.

Please resist catching and handling everything you see. You will never know the amount of stress you have caused.

VENOM or POISON

A rattlesnake is **venomous!**
A dart frog is **poisonous!**
Both words refer to the toxic mixtures of chemicals that reptiles and amphibians possess.
The difference lies in how the toxins are delivered.

Speckled Rattlesnake
Crotalus mitchellii

VENOM IS INJECTED: The rattlesnake and other pit-viper snakes have fangs that work like hypodermic needles to inject the toxin into the prey animal.

POISON IS ABSORBED, INGESTED, OR INHALED: The dart frogs and toads, for example, have glands on their bodies that secrete toxins that, when handled, are absorbed through the skin.

While venoms and poisons can be very damaging to humans, some of the isolated components can also be **very beneficial**. Snakebite anti-venom is made from snake venom.

Cobra venom is used in treating chronic pain. Rattlesnake venom is used in the treatment of high blood pressure and preventing blood clots. An enzyme found in Gila monster venom is now synthetically produced and used in treating type 2 diabetes.

The poison from the **Fire-bellied toad, Bombina orientalis,** a popular reptile in the pet trade, is used in drugs to help identify prostate cancer and for healing wounds.
A full dose of venom delivered by certain rattlesnakes can be lethal to humans if left untreated. Interestingly, rattlesnakes have control over the amount of venom they deliver. The snake may choose not to waste its valuable venom on us and save it for its favorite meal. It is thought that in 20% of all rattlesnake bites, no venom is injected. These are called "dry bites".

The poison from most dart frogs does not kill humans, but if you pick up and handle the **Golden Dart Frog, *Phyllobates terribilis*,** from Colombia, you may die. The poison is strong enough to kill 10 or more people.
The World Health Organization (WHO) reports that 100,000 people die each year from the bite of venomous snakes. Most deaths occur in Africa and Asia. Only 5-10 people in the US die yearly as anti-venom is readily available.
Mosquitos are venomous, and the WHO says 750,000 people die annually from malaria and other mosquito-related diseases.

> ⚠ **IF YOU BITE IT AND YOU DIE, IT'S A POISON!**
> **IF IT BITES YOU AND YOU DIE, IT'S A VENOM!**

CONSERVATION EFFORTS

Around the world wildlife conservation efforts are crucial to helping save all animals that are left on our amazing planet. Trained scientists and citizen scientists are working together to preserve the incredible diversity that we know. A citizen scientist is a person, like you and me, whose interest goes far beyond just watching a documentary on television. With apps like iNaturalist, eBird and a host of others, the citizen scientists report their observations, which helps trained biologists better understand the population dynamics and geographic range of animals and plants.

As for reptiles and amphibians there are many not-for-profit, non-governmental organizations, (NGO's) around the world. PARC is an international organization founded in the UK in 2009. PARC stands for Partners in Amphibian and Reptile Conservation. ARC is a US group, (Amphibian and Reptile Conservancy) that aims at protecting the habitat and the animals alike.

Wherever you live it will be easy enough to find a group of concerned people wanting to preserve and protect amphibians and reptiles. These animals deserve our attention, as they play an integral part in the balance of our natural world.

On a global outlook, I personally find two very important conservation issues. The first is the connectivity of the world's largest natural areas where wildlife exists. The corridors that connect one area to another where animals can pass and breed and expand their gene pool is essential to their long term existence. Yellowstone National Park is a prime example. As big as it is, the large mammals over time will suffer from inbreeding if they cannot move to other wild areas as their populations expand. Establishing wildlife corridors in the United States and in other parts of the world won't be easy, but is essential. Although their movement is slower, even reptiles and amphibians need corridors for their long term health.

Oceans: I could write volumes about how we abuse the waters where much of our oxygen is generated, and where much of the food that feeds the world is found. I hope the young reader, if interested in conservation, will help solve this most important dilemma. We need to protect our oceans for everyone's future.

ACKNOWLEDGEMENTS

So many people have been helpful and encouraging in bringing this project to fruition. Alphabetically, I would like to thank: Erick Berlin of Costa Rica. His friendship and generosity in sharing his knowledge and access to his nature preserve has been a huge part of my life. Daniel Beck, whose work with Gila monsters has set the stage for research and interest in protecting this iconic desert lizard. Jim Boone, the best friend the Mojave Desert has ever had. His lifelong dedication has helped to save endangered environments, and his unique website, BirdandHike.com has inspired thousands of Nevadans to get out and explore. Seth Cohen, a great friend and naturalist who has taught me more about desert reptiles than anyone. Deasminne Cruz, my graphic designer who has done an amazing job. Rob Feiss, an animal lover and friend of a lifetime. Judith Hamilton, for her belief in my project and her encouragement. Josh Hester, my son in law, who has grown to be a desert rat protecting tortoises and the animals in the Mojave. Lonny Holmes for sharing countless hours in the desert and his continual support. Jason Jones, who gave me the opportunity to work with Gila monsters and desert tortoises. He is unselfish with his time, and dedicated to the protection of all reptiles and amphibians. Larry Jones, the "lizard wizard "of the desert Southwest, for his knowledge and dedication to the protection of these incredible reptiles.

Charles LeBuff, for giving me the opportunity to tag sea turtles when I first moved to Florida, and who helped me get started on a long and fascinating journey. Isaac Levine, my project manager from amz publishing for his enthusiasm and staying power in getting this book to print. Kenny Morrison, a fearless and observant citizen naturalist who has taught me more than I can remember. Bernadine Murray whose insight led me to a fabulous experience working with Gila monsters and desert tortoises, and for spending long hours in the often hot and brutal environs of the desert. Danny Perkins, a leader in promoting the use of native plants in Florida. In addition he was always there to help me promote the protection of native animals in the state. Lester Piper, now long deceased, but not forgotten. His Florida zoo gave me the opportunity to work with native animals and meet many wildlife icons. Roger Repp, the Gila monster king of Arizona, whose support I more than appreciate. His sarcasm and wit help keep things fun, but his knowledge of the desert is legendary. Cameron Rognan, for letting me use some of his amazing photographs. Anna Swonetz, who has worked tirelessly on my brand and website development. David Young, for his enduring friendship and tolerance as I often dragged him in the hot and humid environment of South Florida to look for rare plants and reptiles. He was always on board for sea turtle tagging missions on the beach, as long as I had a few beers in the jeep.

Family help was most incredible. Gus, of course, as co-author has been the best. Could and would not have done this without him. My daughter, Nichole, for her perseverance, love and help, in trying to direct me towards sensible marketing, distribution and publishing. My son Joey who has always pushed me to forge ahead. And of course my wife, Pam, who has been cheerleader, editor, critic and unrelenting supporter.

Authors' Bios

James Augustus Hester was born in California and has lived there and in Nevada. Known as Gus, he enjoys snow skiing, golf, art, cooking, and exploring natural habitats. Gus has experience in telemetry tracking Gila monsters and desert tortoises in the Mojave Desert along with his grandfather (the senior author). Gus's drawings of reptiles and amphibians appear in each chapter of the book. He has his own page in each chapter, where he writes about a specific animal. Gus is currently in middle school in Big Bear, California.

Gus measuring a large American alligator skull

James J. Vanas spent 40 years in Naples, Florida where he worked with native mammals, birds and reptiles at a private zoo for five years.
As a citizen scientist, he tagged and protected Loggerhead Sea Turtles on a local beach, and moved nuisance alligators for the game commission. He was an alternate on the first Florida Panther and American Crocodile Recovery Team. Since moving to Nevada he has spent seven years tracking Gila monsters and desert tortoise for the Department of Wildlife. He has traveled extensively through Costa Rica for over 35 years. He enjoys photography, writing and being in natural environments.

Email: jimvanas2@gmail.com

Jim translocating young spectacled caiman from a planned development in Costa Rica

ABOUT THE GRAPHIC DESIGNER

Hi! I'm Deasminne, a designer based in the Philippines and I make meaningful experiences through design. You can visit my life's work here:
deaz-portfolio.webflow.io

www.ingramcontent.com/pod-product-compliance
Lightning Source LLC
Chambersburg PA
CBRC100215040426
42333CB00035B/72